URRAY
Chiltern
Wangaratta
Yackandandah
Beechworth
Gaffney's Creek
Marysville
Woods Point
URNE
Walhalla
AIT
AF580667

OLD GOLD TOWNS OF VICTORIA

The goldfields of Victoria held, in their heyday, the majority of the colony's population. The visible effects of the gold finds of the early 1850s were to be seen most dramatically in the surge of people that they attracted: from 19,300 on the diggings in 1851, to 109,000 in 1855.

This great influx of people, surging and eddying from one rush to another, raised up settlements almost overnight; settlements that were to be as impermanent as the diggings themselves.

Yet a seed of permanence was dropped at most of these early diggings. As the rush faded, some few of the tradesmen, shop-keepers, officials, and innkeepers remained to steady the (as one writer remarked of it) "wild and tumultuous development."

This was evident in the villages that sprang up close to a diggings. Many of these scattered settlements became townships as the variant alluvials gave place to the more ordered mining of reefs. Capital was invested in mining projects, and miners worked for wages.

By the close of the 1850s, the era of the individual digger was closing in the first goldfields. The adventurers still moved rest-lessly from place to place "following the gold"—to the north-east of the colony in the later 1850s, and the mountains and valleys of Gippsland. The wages men began to put down roots, and townships developed to meet their needs.

Many of these townships had fortunes that fluctuated with the rise and the decline of the mines that gave them life. Some survived that decline, to serve the needs of farmers as the lands were thrown open to settlement. Some stood still, or slowly decayed. Others disappeared, and a remnant continued a shambling existence.

There were exceptions, both as to the degree of survival from the age of gold, and as to the growth that gave a more lasting stimulus than the gold upon which they were founded. Ballarat, Bendigo and Castlemaine are examples. They have long since outlived the mining period; their growth from other reasons makes that period appear as transient.

The gold that established these and many other centres has gone. But all of them retain some relic of that period; all of them have a heritage of glamour, memory, or mining remnant.

This is coming to be increasingly recog-nised by citizens of the old gold towns, by people who have found rest and recreation in them, by historians, architects, antiquarians, and conservators.

A re-discovery of these old towns is now taking place. Some attract visitors by their positive glamour. Others have inherent his-torical interest. Some demonstrate particular aspects of survival, or have a special holiday or recreational appeal or a definite place in our national and cultural heritage.

Many are proving pleasant havens for city dwellers seeking relief from stress and strain. The motorcar is bringing new life to old, perhaps ghostly, gold towns. They are attractive refuges during weekends and

BEECHWORTH Old Hotel, Camp Street

holidays, in which people become fossickers for gold and gems.

All this is giving impetus to the urge to preserve and to renovate old things. It is making us take a more mature and objective look at our old gold towns, at the survival in them of treasures that have been battered by neglect; it is giving them a new place in the story of our past.

Encouraging support has come from local historical societies, dedicated citizens, and the National Trust. The latter is awakening wider public interest in the need to preserve our heritage, and to re-create the characteristics of our old and long neglected gold towns. There are some exciting examples of this new interest, and of concerted moves that are being made to capitalise on it.

Thus, Maldon is to keep a living place in our gold era history as the first Notable Town in Australia. We have come to realise that this old town, because of the variety of its building styles, is important to students, historians, and antiquarians. There is no architectural significance in its street of verandahed shops winding up the hill, but it shows the character of the town which it is so important to retain.

The Government of Victoria has recognised the unique value of Maldon in our history by adopting the Trust's classification of the town.

Beechworth is another old gold town that is being revived by the great interest that people are taking in its place in our goldfields heritage. This centres mainly on its notable buildings, rather than on the town as a whole. Beechworth is fortunately placed, in one of Victoria's outstanding scenic districts, and was the centre of a most glamorous gold mining area. Much of this glamour is retained in the town itself.

Walhalla, another of the mountains goldfields, could also be revived as a tourist and holiday centre. A few landmarks remain of the once lusty town that was built, clinging to a mountain side, by gold. The Walhalla Improvement League, and its supporters, believe that a revival project can be built upon these remnants. They argue that gold and ghosts can give the old town new life as a tourist resort.

The large cities, such as Ballarat and Bendigo, have long since looked away from gold. But they have not lost interest in preservation, and re-creation. A mining village is taking shape in Ballarat. All the machinery and plant of a mine at Bendigo has been bought by that city, and is to be made into a tourist attraction. In such ways, the history of the goldfields is being preserved in tangible form.

The paintings in this book have set out to say something about points of interest in relation to the past of some old gold towns. They do not pretend to give an overall assessment, but represent the attitude of the artist to an old town, and something in it that he felt should be captured.

The notes accompanying the paintings are intended only as impressions of some towns that played a part in the heroic years of Victoria's golden past.

CRESWICK The Wide Road Through

Gold was first found at Creswick, which is only ten miles north of Ballarat, in September 1851. The site, known originally as Creswick's Creek, took its name from Charles, John, and Henry Creswick, who squatted there with sheep in 1842.

A great rush occurred in 1852, when rich alluvial yields were obtained from shallow workings. There was another spectacular rush in 1856-1857, when deeper workings were opened up.

By 1861 it was a stable town and centre with a population of 5,000. The alluvial claims were important for many years. Even when Ballarat suffered a decline in the 1870s, alluvial yields from them were steadily maintained and gave work to 4,000 miners.

The Australian mine was the scene of a terrible disaster in December 1882, when it was flooded by water breaking in from old workings. Forty-four men were caught at the working face, and twenty-two drowned and suffocated. The hymns sung by the trapped miners, before water engulfed them, were the essence of the hopeful courage of humble men.

Large areas of the old mine workings are now covered with pine plantations of the State Forests Commission on the fringe of which is the School of Forestry.

Light industries such as textiles, forestry products, and case making are important now; district industries include grazing and mixed farming. It is an attractive holiday centre.

BENDIGO Rifle Brigade Hotel

Was a woman the first to find gold at the Rocks on Bendigo's creek? Margaret Kennedy, wife of a Ravenswood overseer, who had brought gold specimens from Buninyong, was one of thirteen claimants in 1891, when a special committee considered making a reward for the find. None was made; the evidence was too confusing.

Bendigo was the nickname of a shepherd, after an English prize fighter. The first gold was found, in late September or early October, 1851, near Bendigo's hut on Bendigo's creek. The party seen washing gold there included Margaret Kennedy and the wife of Patrick Farrell, a cooper.

News of the find was broken to the people of Melbourne in the *Argus,* on 13 December 1851, and the rush turned the green, wooded valley of Bendigo into a treeless, slushy swamp.

Bendigo passed its peak of alluvial production in September 1852, but, in reduced numbers, the diggers stayed on. The amplitude of its golden wealth was not then dreamed of. That came from the exploitation of quartz reefs, and great, deep mines spread over thirty-seven district lines of reef, in an area of about twenty miles by seven miles.

The nucleus of Bendigo township (then Sandhurst) was below the government camp on Camp Hill; a rise above the town site of 70 acres. The first stores, buildings and dwellings were erected along the line of a track at the foot of the hill. This area is the attractive heart of Bendigo today.

MALDON Wintertime in Maldon

BALLARAT Mair Street

MALDON View across High Street

Maldon seems destined to spearhead the visible belief in preservation of the heritage of Victoria's golden years. By an official order issued in January 1970, much of the shopping centre of the town, and several buildings of historical importance, are to be preserved intact.

The town, eleven miles from Castlemaine, is the heart of one of Victoria's most interesting old gold mining districts. Gold was first found there in June 1853, by a party of nineteen prospectors financed by John G. Mechosk, a German. He spent $3,800 on wages and rations for the party. The find caused a rush to the Tarrangower ranges, to be described by William Howitt later that year as "man thronging on man."

Gold laid Maldon's foundation and contributed to its development for over seventy years. It survives placidly as the centre of wool, wheat, oats and barley growing; some timber-getting, and an increasing tourist traffic attracted by the unique character, and historical value, of its buildings.

These have no architectural significance. Their antiquarian and historical interest lie in the variety of primitive design and construction that have survived from the town's first years. The cottages and outbuildings are in a variety of styles; wattle and daub, horizontal and vertical timber slabs, pisé and mud brick, weatherboard, brick and stone. The verandahed shops of the main street retain the character of a town that once bustled with prosperity.

MOLIAGUL Gold Diggers' Pub

Moliagul, a small, old town, is truly a place of nuggets, in keeping with several old diggings nearby.

On 5 February 1869 two Cornishmen, John Deason and Richard Oates, who were fossicking on a slope of the Bulldog Gully, came upon a large nugget that was almost pure gold. The great mass rested only a few inches below the surface. It was twenty-one inches long and ten inches thick; the largest piece of gold of its kind ever known.

The finders carried it in excited triumph into Moliagul, breaking off scraps for friends along the way. It was broken into three pieces for easy carrying and taken to a bank in Dunolly nine miles away. The bank manager bought it for £9,436.

The nugget was called the Welcome Stranger. It weighed 2,520 ounces, and yielded 2,260 ounces of gold when smelted.

An obelisk in the bush stands near the spot where the nugget was found.

The town of Moliagul (present population about seventy) now thinks little of that great mass of gold. It may even be prouder of John Flynn, who founded the Flying Doctor service in 1928, for he was born in Moliagul on 25 November 1880. A memorial in the local recreation reserve marks this notable event in the town's story.

EAGLEHAWK Relics of the Past

Eaglehawk Gully, five miles from Bendigo, was perhaps the most famous of the rich alluvial diggings on that field. The first finds were made there during May 1852. They caused a rush of 8,000 diggers, who found gold glittering in the surface dirt of a wide area of scrubby ground.

These superficial deposits yielded dazzling rewards to men who scrambled over the ground on the Red Hill (later Virginia Hill) and scooped surface shreddings from the quartz reefs into tin dishes. The whole face of the hill was covered to a depth of from six inches to two feet with these fabulously rich shreddings.

It was said at the time that individual diggers and small parties lay over their claims to prevent neighbors from scooping the earth into their own claims.

This richness kept thousands of men on the field for over a year, after which no more than a thousand remained. They were still getting returns from the surface shows.

Later, quartz reefs were exploited to great depths. Evidence of the heyday of Eaglehawk's mining industry is to be seen in sand and mullock heaps, old engine houses and mine workings. Attempts to revive mining have not succeeded, and other forms of industry have not been supported.

Many of Eaglehawk's buildings and houses survive, together with new ones that are making it a residential area.

WALHALLA The Old Bakery

Walhalla is Victoria's most romantic old gold town. Its remains are spread along the mountainsides on the banks of Stringer's Creek, and in its heyday the town boasted three breweries, fourteen hotels, and 5,000 people. Today there are forty houses and thirty-two people. The Walhalla Improvement League hopes to revive the place as a popular holiday resort; an ambition which is approved by the National Trust.

It has the setting. The road through it is as steep as the once populous streets; can the dwellings of the people who once lived there be recreated, and the once glamorous atmosphere revived?

Anthony Trollope, the English writer, stayed a night there in the summer of 1872. He marvelled that there was a piano and a billiards table in the town. He wrote that: "The roads were so steep it was often impossible to sit on horseback . . . I could not have believed there had been so much traffic across the mountains and through the forests."

Yet, from the early 1860s, prospectors found rich reefs and pockets of alluvial gold at Stringer's Creek and along the lower Goulburn River. They founded the town, which in its heyday was a single main street. From twenty to forty feet wide, it competed with the stores, banks, offices, mine batteries and public institutions for the limited level space along the creek banks. The miners' dwellings were perched on the steep slopes above the town.

The field was handsomely rich. The Long Tunnel company, opened in 1863, returned £510 for each original £10 share during its forty-five years of production.

BENDIGO McCrae Street

WALHALLA The Old Brick Mill

TARNAGULLA The Old Blacksmith's

When his barque was wrecked in Poverty Bay, New Zealand, Captain David Hatt was rescued by a Maori girl named Margaret. He married her and brought her to Victoria, settling at Sandy Creek. Here, fossickers had found nuggety gold in the alluvials, but Captain Hatt and a small party found the main lead. Though it was rich with gold, he named it Poverty Reef in memory of his shipwreck. The area, near the Loddon River, was rushed in 1853, and 5,000 diggers worked claims for two-and-a-half miles along the creek.

The town that grew there was named Tarnagulla.

More than 170 quartz reefs were worked in the area, many of them high yielders. The Poverty Mine yielded more than thirteen tons of gold before it closed in 1894. Some of the early crushings, it is said, gave more than 200 ounces to the ton.

There was a revival in 1906 when an old miner, Jack Porter, opened up a reef about four miles out of the town, then a thriving farming centre. On Melbourne Cup day, Porter bottomed a claim at eight feet and found a nugget that weighed 593 ounces. It was the last of the great nuggets found in the district. The mine's name was ready made: Poseidon, which won the Cup that day.

Soon, the whole countryside was pegged. The rich lead ran down to the Loddon, but flooding, and the outbreak of war in 1914, put an end to the Poseidon.

GORDON The Old Pub at Gordon

Gordon, once a busy roadside township, is now little more than a bend in the highway to Ballarat. It was named after George Gordon, who in 1838 occupied a stock run of 30,000 acres. Charles Egerton, a neighbor to the south, held 35,000 acres.

During 1853, gold was found by Alexander Russell in a gully on Egerton's run, later known as All Nations' Gully. At the crest of Mount Egerton a reef was found, later celebrated as the Mount Egerton mine. Thomas and Somerville Learmonth paid £7,000 for the mine in 1863. They received small returns, and on advice of their mine manager, William Bailey, sold to a Ballarat speculator for £13,500.

The mine flourished immediately and the Learmonths sued the new owners and Bailey, alleging that he had allowed the mine to run down. They failed after costly litigation. The mine was worked until 1910, produced gold valued at £2 million, and paid £493,000 in dividends. Bailey built a mansion, now the St John of God Hospital, in Ballarat.

Gordon township profited from the Egerton gold riches. Robert Evans (known as Kangaroo Bob) found gold there in 1853. Soon afterwards, a bullock driver named Parkes opened up a reef north of the township. Both mines were developed, as the Kangaroo Bob and Parkes United, and were high yielders.

Gordon developed from a scattered hamlet of slab huts and tents to a township with banks, stores, and eight hotels. One hotel remains. Commercial mining ceased in about 1910.

KYNETON Country Pub

When Kyneton was surveyed in 1849 its only buildings were a small homestead at St Agnes run, taken up by the Wedge brothers in 1838; a slab police station and lockup; and a couple of huts. Even then it was a stopping-over place for teamsters, carrying provisions to the stations and wool back to Melbourne. They called it Campaspe Flat.

The bullock drays made the winding track that became Kyneton's main street, when the gold rushes to Bendigo and Castlemaine filled the line with houses, huts, tents, shops and wayside inns.

Gold gave the township great prosperity, as a reservoir of foodstuffs and supplies, and brought bustling life to shack towns along the goldfields road. In the mid 1850s, a Kyneton publican advertised stabling for 1,000 horses.

Not much gold was found in the vicinity in the early gold years, but by 1853 the township wore an air of permanence, with fine bluestone stores, hotels, banks, and dwelling houses; there was a wooden church and a stone parsonage.

In the 1880s, gold was being profitably mined at shallow depths on reefs nearby, and at Taradale, Malmsbury and Lauriston. Then, the Kyneton mining exchange had fifty members.

It is a picturesque town, snugly situated in a wide bend of the Campaspe river, the centre of a rich mixed farming and lamb raising district.

DAYLESFORD The Old Hillside Store
(by kind permission of Major & Mrs J. Ball)

Daylesford is a picturesque town with substantial public and commercial buildings, sited on a hilly spur of the Dividing Range and near the timbered Wombat Creek.

Gold was first found there in August 1851 by two prospectors, J. Egan and T. Connell. They were rewarded with £800 for finding a new field, then known as the Jim Crow diggings.

Early rushes did not yield great returns, but the town grew steadily from the work of diggers along the creeks and gullies. In 1856-1857 a more spectacular rush brought 5,000 diggers onto the field. A substantial mining industry developed, especially in the 1880s.

Daylesford's civic and commercial growth kept pace with mining. This is to be seen in many substantial buildings that have survived, an elaborate town hall, banks, large stores and churches. The high school and technical school are more recent examples.

Interesting mountain scenery, mineral springs at nearby Hepburn, and two fine lakes make the town a prosperous holiday centre. Potatoes, wool, dairying and sawmilling are the chief district industries. There is still some fossicking in the creeks, holiday-makers being the most enthusiastic.

CASTLEMAINE Beck's Imperial Hotel

Four workers on Ravenswood sheep run, inspired by Ballarat and Clunes, washed creek gravel in the granite shadow of Mount Alexander. On 29 July 1851 they washed fifty pounds of gold, and wrote to the Melbourne *Argus.* The resulting rush brought 20,000 diggers by mid-December; most of Victoria's gold-searchers. They swarmed over fifteen square miles of wealth-yielding creeks, valleys, and flats, cut down the forests, ripped the soil down to bedrock with their dredges and sluices. La Trobe wrote to the Colonial Office: "A pound weight of gold a day is small remuneration for a party," but the scars of their depredations may never be effaced.

Castlemaine now stands where the first government camp was established, in 1852. Ten years later, a "solid, secure, and surprisingly modern-looking town," was growing there. By that time, the most impressive and important buildings had been erected. They still stand, as memorials to the turbulent goldfield days.

Nearby villages and hamlets linger on from those glamorous times, but the pleasant city of Castlemaine now thrives on well-founded industries and agricultural activities. But it is conscious of its golden past, and is striving to preserve the historical heritage which attracts visitors to the city and its environs.

RHEOLA Melville's Caves

Legend has it that Captain Melville, the bushranger, used the caves in the mass of granite near the old gold diggings of Rheola as a hideout.

If that is so, it was before gold was found in the district. Because Frank McCallum, who called himself Captain Melville, was captured with a mate in Geelong on Christmas Eve 1852; Rheola's gold rushes occurred later in the 1850s.

McCallum was sentenced to thirty-two years in gaol, and sent to the hulks in Hobson's Bay. He was one of a boat party that attacked a guard in October 1856. Sentenced to death, he was reprieved. In July 1857 he killed the governor of Melbourne gaol. A fortnight afterwards he was found dead in his cell.

The caves are now a national park reserve, and form one of the natural curiosities of the district.

The Rheola diggings were noted for the many large nuggets found there. Ninety-eight notable nuggets were recorded, with weights up to 1,100 ounces.

It is a small township on Kangderaar creek, eleven miles from Inglewood, in a district of general farming and fruit growing.

TARNAGULLA Country Street Scene

CASTLEMAINE Vista Down the Road

BUNINYONG The Elegant Old Store

In August 1837, Thomas Learmonth and a party of six landhunters from Geelong viewed the timbered lands north from the crest of Mount Bonan Yowang. Two years later he settled there with sheep and the name was changed to Buninyong. A township was started soon after by sawyers working in the thick timber of the gullies. It became a substantial town, with hotels, a school, and a church, and was an important stopping place for traffic from Geelong.

A timber getter, Thomas Hiscock, was first to find traces of gold, in a gully that bears his name. This was on 3 August 1851, only a few weeks after Clunes gold was discovered. It was the first important gold find in Victoria, and men rushed it from Clunes. At first returns were small, as they knew little about the techniques of washing for gold. It was a bitter winter, with snow on the high hills and floods in the valleys.

By that time, Buninyong had grown to be one of the largest inland towns in Victoria, and had a Church boarding school, more hotels, a doctor, blacksmith, and storekeepers.

Hiscock's find led to the opening up of Ballarat a few weeks afterwards; through Buninyong were to pass thousands of diggers on their way to the rich alluvial diggings six miles to the north.

Now, Buninyong is the centre of a flourishing farming district, a quiet township on a busy highway.

HEATHCOTE The Old White Store

Heathcote had a brief period of glory, as centre of the rich McIvor diggings. Gold was first found in December 1852 on the McIvor Creek at the foot of Mount Ida, in rough, timbered country.

Within five months, it was said, 30,000 diggers flocked to the field, and, characteristically, a substantial town was established. The diggers had mostly come across the granite hills from the gold road at Kyneton. Few remained for long.

The known yield of gold on the McIvor to June 1853 was 21,458 ounces. A decade later the field was virtually deserted, although the town had spread for a considerable distance along the highway. Its main street was four miles long from north to south.

On 20 July 1853 the gold escort from the field to Kyneton was held up by six men, who had felled trees across the track in thick bush country. Shots were fired at the escort of two officers, three troopers and the cart driver. The shots scattered the escort, and the bushrangers got clear away with 2,000 ounces of gold and £700 in cash.

Five of the men were caught, and three gaoled for life. One turned informer and later suicided; the fifth gave Queen's Evidence. Most of the gold was recovered.

BALLARAT Symmetry in Iron Lace, Lydiard Street

From 1838, when Ballarat was first occupied as a sheep run, until August 1851 when gold was first found there, only seventy people lived on the swampy banks of Lake Wendouree.

A few days after the find, 400 people were on the field. The *Argus* said: "Gentlemen (are) foaming at the mouth, ladies fainting, and children throwing somersets, and all on account of the extraordinary news."

Little came of these first finds, but when good finds were made at shallow depths in October 1851 there was a rush of 10,000. This was doubled by 1853, when 319,154 ounces of gold were taken by escort to Melbourne.

Ballarat was probably the richest alluvial field the world has known. Big nuggets were found during the 1850s, culminating in the Welcome Nugget, of 2,217 ounces, found at Bakery Hill in June 1858.

The alluvials faded early, but from the early 1860s the quartz leads were worked. The field was at its zenith in 1868, with an estimated population of 64,000 and 300 mining companies. A recession in 1870 caused thousands to leave. Many mines closed and investors lost confidence, yet the mines were still high yielders.

Gradually, the city recovered. Woollen mills and iron foundries were opened, and the deep mines were worked, although in smaller numbers. The total production of gold to 1954 was 20,135,000 ounces, second only to Bendigo in Victoria.

The modern Ballarat is built over the "monarch of goldfields" as it was described in 1858, and is an important, flourishing city. Its 42,000 people depend on vigorous secondary industries, and it is surrounded by well-developed farms and pastoral properties. The parks, gardens, and recreation grounds for which twenty percent of the city area is reserved, make it a beautiful city.

BEECHWORTH Ford Street

Beechworth is a town set amidst some of Victoria's most beautiful landscapes. It is the centre of a ready-made tourist area, and a vivid reminder of its romantic heyday as a gold diggings.

In 1966 the National Trust decided to attempt preservation of the town's historical sections as a living entity. It was a mammoth conception, undertaken with some misgivings. But it is now apparent that given time it will succeed, establishing the accuracy of the Trust's conviction that progress and preservation do work together "for the community's benefit."

When a shepherd named Howell first found gold in Spring Creek, the only building was a bark hut. Soon, thousands of diggers stormed in, and in the first four months of 1853 the gold escort took 123,000 ounces out of Beechworth. In 1857 a visitor recorded that it was "the thriving centre of a busy, prosperous community numbering close on to 20,000 souls with numerous churches, banks, great stores and spacious hotels, boasting also a splendid hospital."

The gold ran out at the turn of the century, but everything in the town was basically intact, so that now its great asset is that with careful preservation and restoration, there is a treasure of the golden era. The district's chief products are hops, tobacco, apples, nuts, maize, potatoes, and timber. Its chief import is tourists, seeking a past glory.

MALMSBURY Hillside Shops

Malmsbury is an old bluestone town in the valley of the Coliban. Most of its buildings have a neglected air; Town hall, Mechanics Institute, churches, two-storeyed shops and houses, an old stone flour mill in decay.

Once, it was a busy railway centre, which accounts for its large station buildings, goods sheds and yards. They were needed when the town was a busy transhipment point for the inland gold centres, for superb local freestone and bluestone, for wheat, oats and hay, for flour gristed at two steam-driven mills, and for timber.

By 1860 there were twenty-eight hotels and drinking shanties in the town, and two breweries. Today there is one hotel.

Gold, both alluvial and quartz, was won from shallow workings in the 1860s. This helped to sustain the town after the railway continued through the town in 1862; a heavy blow (a local historian noted) "at the prosperity of Malmsbury," because it was no longer the rail-head.

The main deep lead was never found, although bores put down sixty years ago proved that it ran for more than thirty miles through the Kyneton-Malmsbury district.

Malmsbury remains an attractive old town, with treelined streets and pleasant picnic spots where diggers and bullock teamsters rested beside the river in the golden 1850s.

WOODS POINT The Mountain Way

Gold mining was depressed in the 1860s, but a great revival came from rich finds in the little-explored mountains, valleys and streams of remote Gippsland.

William Gooley, a lone prospector, found coarse gold in a creek downstream from the present-day small township of Woods Point, in May 1861. He pegged three claims, but when he returned from Jamieson after registering them, they had been jumped.

He got them back by his own efforts, and never revealed how much gold he won. When asked by a gold reward committee how he had lost an eye he replied it was "in the just defence of my rights." He was awarded £100 for the find.

The name Woods Point was first used during April 1863, after an American, Mabelle Woods, the first storekeeper of the township, which was perched on spurs in a valley. Woods went prospecting one day, and never returned.

Two men of Oban, Scotland, Duncan and Colin McDougall, developed the rich quartz depoits. They set up a water-powered battery, the parts for which were packed in by horses, a heroic three months' trek. They crushed £6,545 worth of gold in August 1862; it was hidden in the scrub and carried away secretly at night. This claim became the Morning Star mine.

The field was sensational for its reefs, its boomed claims of no account, and the longevity of its great mines: the Morning Star until the middle 1960s, and the Al Consolidated which is still working.

GAFFNEY'S CREEK Mountain Store

Small parties and large numbers of individuals prospected the mountain reefs of Gippsland from Walhalla to Woods Point, and further in; sixty miles through dense forests, along brawling creeks and steep valleys. They camped on precipitous hillsides, to which they had packed their provisions and mining equipment on horseback.

A few of these hardy men made fortunes; some a competence. Most of them worked the reefs and the alluvials with no reward, yet continued to look for gold. Some, whose fate is unknown, did not come out of the mountains.

One of these men who followed the gold traces because it was in his Irish blood and not for any riches, was Terence Gaffney. In January 1860, he worked a lonely way up the Goulburn river to the lower reaches of the creek that now bears his name. He found gold. A rush of 400 diggers followed him into the mountain fastnesses.

A township was established on a spur, with a public house, stores and butchers' shops in temporary premises of logs and canvas.

The track into Gaffneys in its early days was so precipitous that pack horses had to be left to pick their own way up and down. At the height of the boom in late 1861, 500 horses and mules carried supplies and machinery to the field from Jamieson, which was the gateway to the district and the chief supply depot.

STEIGLITZ The Derelict

Steiglitz is indeed a ghost town of golden memory. Its remains stand fifty miles west of Melbourne between Geelong and Ballan. A century ago its population was 5,000, in a town of substantial stone buildings busily occupied: court house, church, stores, school, houses, powder magazine. Today, with an estimated permanent population of ten, only remnants remain.

The town is named after the Steiglitz family, pioneer squatters of the district. Its pastoral peace was disturbed during 1853 when a party camped at a spot near the junction of the Graham and Sutherlands creeks, and noticed gold streaks in quartz stone.

A year or so later, seven hundredweight of this stone was crushed in Geelong and yielded fifty ounces of gold. Within ten days there were 200 diggers along the creeks. In 1856, three crushing machines operated, and there was alluvial mining in the creeks.

A substantial town arose with the peak of mining activities. In 1894, 2,000 people were working the reefs, but by 1896, shops were being advertised for removal. By 1914, the population was only 500.

There has been no recovery. The old powder magazine was bulldozed some time ago, and some old mine workings were filled up. They were part of the seventy-five quartz reefs which were scattered over fifty square miles.

TALBOT The Tall Town Hall of Talbot
(by kind permission of Mr S. Gibson)

Talbot, then named Back Creek, was only three miles from the gold discovery at Amherst during 1852. Talbot's own rush did not start until 1859, when rich alluvials were found in the Back Creek.

The richest of the leads, the Scandinavian, was found that year. It was under what is now Talbot's main street. That find caused the first large rush of an estimated 50,000 diggers and the usual camp-followers, to cater for their needs and entertainments. In the 1860s there were 35 registered hotels on the Back Creek diggings, and twice as many grog shanties.

The name was changed to Talbot in 1861, and in 1865 when it was proclaimed a Shire the present fine Town Hall was erected. The town (population 436) now serves a rich mixed farming district.

It is an attractive old town, popular with holidaymakers, who hopefully fossick in its creeks for the gold that made it a much larger centre in the 1860s.

WEDDERBURN Corner Store

Gold diggings were opened near Wedderburn, then known as Mount Korong, in September 1852 because of the acumen of John Hunter Kerr. He was a local squatter who encouraged his shepherds to look for gold, which was then attracting thousands to Bendigo and other central Victoria fields.

The find was rushed and some claims gave high yields from shallow workings.

Mostly the field was nuggety, and several large pieces were found. The first of these was the Blanch Barkly nugget, which was unearthed near Kingower during 1857. It weighed 1,743 ounces. Another large piece was the Viscount Canterbury nugget which weighed ninety-three pounds three ounces and yielded 1,114 ounces of pure gold.

There were several prosperous fields in the Korong Shire, of which Wedderburn is the administrative centre, but some rich reefs in the area were hardly touched. There are still some hopes that they may yet be worked.

One such hope of revival of the industry was the finding during 1950 of two fairly large pieces of gold on the surface in the town itself: the Wedderburn Dog of 145 ounces, and the Golden Retriever of 72 ounces. They were both found in a small auriferous area that unaccountably was not worked over during early rushes.

INGLEWOOD Country Township

Inglewood is the municipal centre of Korong Shire, now a flourishing agricultural and pastoral area. In the 1860s at least half a dozen rich gold diggings radiated from the town, amongst them Kingower, Rheola and Wedderburn.

All of the Korong diggings were noted for nuggets that were unearthed from just below the surface.

The first important find in Inglewood itself was made by J. Honey and the three brothers A., F., and J. Thompson in November 1859, the year of many wild rushes in the area. The rich find was made in Psalm Singers' Gully, and the rush that followed drained miners from as far away as Castlemaine. An estimated 30,000 diggers were on the field by 1860, but a year later this number fell away to 2,000. This was the nucleus of the town's settled population, now serving a steadily developed farming district.

The field, like some others of the period, was comparatively unproductive, and on the perimeter of earlier rich development. Yet, as late as 1910, Inglewood was described in a municipal directory as: "a mining township in the centre of a splendid agricultural and pastoral district."

CLUNES Shades of Past Glory
(by kind permission of Mr W. A. Nette)

Clunes township is situated prettily in a valley of the Deep Creek, ninety-seven miles by rail north-west from Melbourne and twenty miles north of Ballarat. It was described recently as: "a homely little community that the small-time farmer shows no desire to leave . . . smaller, less exciting, yet far more stable than it was a century ago."

It has more amenities than its population of 800 would seem to warrant; halls, a modern district hospital, and all sporting facilities. A rich agricultural and pastoral district is the town's main support.

James Esmond was first to disturb the pastoral quiet when on 28 June 1851 he found gold in quartz reefs along the Deep Creek, near the homestead of Donald Cameron's Clunes sheep run. It was later recognised as Victoria's first commercial gold field, and Esmond was rewarded with £1,000 for the find.

Only fifty people were at the Deep Creek by the end of July 1851. They panned dishes of dirt in the creek and scratched at the seams in the wide quartz reefs, but their labours barely paid.

Rich alluvial leads were found when the Port Phillip and Colonial Mining Company brought machinery onto the field. By 1857 there were 1,500 on Clunes field, and the reefs had produced £88,000 worth of gold.

In the early 1860s there was a floating population of up to 40,000. By 1873 the town had forty licensed hotels and many grog shanties, an exciting mining field of great riches, and several small industries. A knitting mills is the chief employer today the town's main support is a productive agricultural and pastoral district.

CHILTERN Old Shops

Chiltern, 170 miles north-east from Melbourne on the Melbourne-Sydney road, started like so many of our early towns as an inn; the Black Dog, on the creek of that name.

It was a humble wooden structure put up in 1844, and became a changing station for the mail coaches. A brick hotel was built in 1851 and named the Horse and Jockey. A police camp adjoined it on land leased from the publican.

Black Dog township was approved in 1853, and the name changed to Chiltern in 1854.

Gold was first found in 1858, attracting diggers in large numbers. A rush at Indigo Creek in October was caused when an innkeeper, F. H. Morgan, reported rich gold found by a party he employed. The diggers found none and threatened to lynch him. A committee of four went down the shaft, and reported to the lynching party: "It's alright boys; there's gold there; get your pegs in as quickly as you can."

The rush saved Morgan's life and built a township almost overnight. There were 10,000 in the district by the end of 1859. A new township site was built on and proclaimed in 1861. Over many years until about 1928 gold mining was a flourishing industry, and the town important commercially. Its population is steady at about 1,500, with district industries including timber, mixed farming and wine-growing.

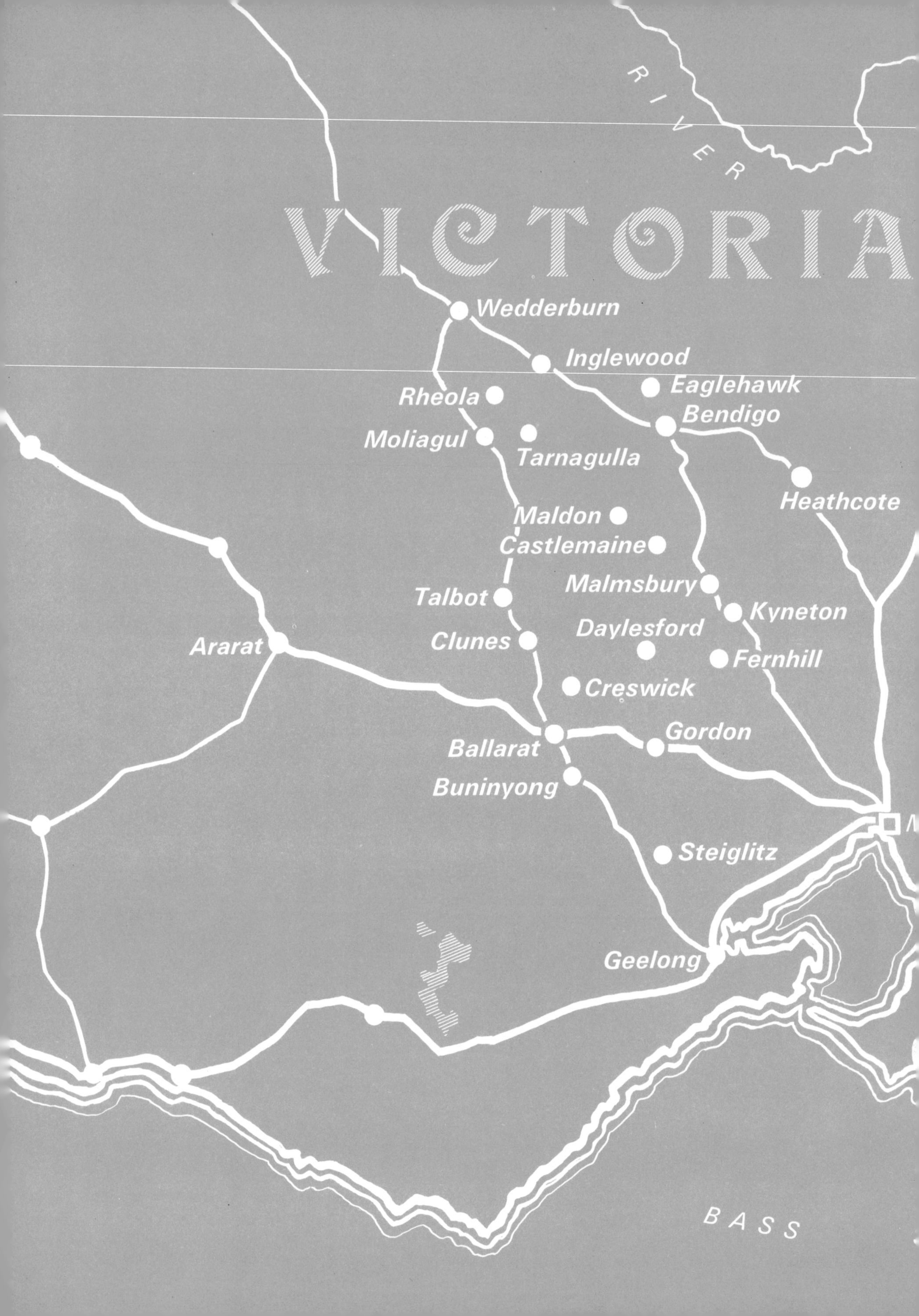

RIVER

VICTORIA

Wedderburn
Inglewood
Eaglehawk
Rheola
Bendigo
Moliagul
Tarnagulla
Heathcote
Maldon
Castlemaine
Malmsbury
Talbot
Kyneton
Daylesford
Ararat
Clunes
Fernhill
Creswick
Gordon
Ballarat
Buninyong
Steiglitz
Geelong
BASS